Podcasting
The Companion Handbook

A Guide to Producing and Publishing Your Podcast

Steven Christianson

Henley Point
Toronto

Cover design by Josée Scalabrini and Henley Point Productions
Interior map image, with permission, by Vecteezy.com
ISBN: 978-1-7778347-1-5

CONTENTS

ABOUT *THE COMPANION HANDBOOK*

Podcasting - The Companion Handbook: A Guide to Producing and Publishing Your Podcast is a workbook, guide and templates for planning and production, a reference guide to terminology and key issues in the world of podcasting, and a production scheduler for podcasters with any level of experience. *Podcasting – The Companion Handbook* is a compact, portable accompaniment to *Podcasting: Explained!*, and guides the podcaster along the practical steps in developing, publishing and maintaining a podcast.

Podcasting – Explained!, published in 2021, is the "parent book" to the *Companion*. It is a larger treatment of how podcasting fits into the world of content creation and promotion. *Podcasting – Explained!* shows how this medium can be used to its fullest, as well as podcasting's ongoing relationship with radio broadcasting, and how the world of podcasting might evolve.

Get your copy of *Podcasting – Explained!* at an Amazon Global Marketplace near you.

WHAT IS A PODCAST?

Podcasts are conceptually very interesting creatures. They constitute content as well as a medium in and of themselves. Put very simply, the podcast is a digital audio file made available on the internet for downloading to a computer or mobile device. They are often available as a series, where new installments of which can be received by subscribers automatically. They are not fully subject to broadcast rules and regulations. They can be as short or as long as desired. They can feature one or multiple hosts. Upload and distribution can be as frequent as the author chooses. The author, in fact, is referred to as the podcaster.

PODCAST FORMATS

The format, or style, for podcasts tends to fall into one (or a combination) of the following.

The Solo Commentary/Narrative

This format typically features a single podcaster speaking, and delivering an audio commentary about some issue or topic. The solo commentary or narrative is much like an "opinion piece", and running times tend to range from three minutes to twelve minutes.

News Read/Current Affairs Delivery

This format is similar to the Solo Commentary or Narrative with the exception of not including opinion or commentary. This format delivers a "reading" of the news. To the listener, this content sounds like a news program. Individual pieces are often combined thematically into a single episode, thus running at approximately four to five minutes.

Documentary

The documentary style provides an excellent accompaniment to more central content. For example, whereas the central content might focus on an episode that features a Current Affairs delivery, the complementary podcast to this "news read" could offer more research and background material for the listener in a stand-alone episode. Some podcast programs focus solely on episodes that are, in essence, audio documentaries. Traditional news radio, such as CBC, NPR, or the BBC, tends to produce a healthy inventory of such podcast offerings. News platforms like *The Economist* also feature plenty of documentary-format podcasts and podcast episodes.

Interview

The interview format is, by far, the most popular podcast format streamed today. Whether in-studio or remote, this format features a host interviewing a guest whose work or experiences reflect the main content of the host's episode or entire podcast program. Some are short, some are scripted and many are free-flowing and entirely unedited. Interview formats vary considerably in running time.

Co-Host Discussion/Dialogue Exchange

This format is also among the most popular for podcasters. Generally speaking, this arrangement features two people with interests chatting about some specific facet or issue related to their interests. Sports, cooking, entertainment, travel – the topic of focus is endless. This format also includes the less-focused co-host discussion, one in which two friends begin recording and chatting about whatever strikes their fancy, in a largely unscripted, unedited and meandering exchange of views and opinions.

Fiction/Serial

This type of format is similar to a book, in that each episode represents a chapter (or, at least, part of a chapter). This format of podcast is growing in popularity. Not dissimilar to an audio book or radio serial, this format generally requires a well-scripted outline, and a final product that has a polished edit.

Video

The video podcast is, at the moment, still in its infancy – if only due to the fact that only a handful of streaming apps and platforms offer users the ability to enjoy the full video podcast experience. This format can entail a live video capture of the studio experience, which is the format chosen by most video podcasters; alternatively, the video podcast is much like any video content one might view on YouTube.

Second Source Content (speeches, conference panel discussions or lectures re-packaged into podcast episodes)

Provided someone has the appropriate permissions and copyrights, this option offers a format that delivers "packaged" content by other people. Educational institutions often use this format to capture, upload and distribute podcast content of teaching instruction. This format is also used quite effectively as "accompanying content" for a seminar or conference, an audio version of "related" or "further reading".

Corporate Messaging (such as newsletters)

Company podcasts are slowly evolving, but, to date, have yet to tap into their true potential as a form of content creation and promotion. Most companies produce a newsletter of some sort. The Corporate Messaging podcast often features audio podcast versions, or customized adjunct material, available only to employees or exclusively through subscription. These formats can feature any number of styles. They are also quite effective for hybrid, remote and international working staff. Direct messaging from corporate leadership (be it government, military, political or religious) is packaged into audio content that complements the existing inventory of outreach materials.

Mirrors (such as radio programs published as "podcasts" with little/ no change to the original radio format or style)

In the parent book, *Podcasting – Explained!*, both podcast "mirrors" and podcast "sprouts" are discussed (the "Terminology" section in this book also serves as a useful reference). Both formats tend to cluster in the world of traditional print and broadcast media. A typical example of a podcast mirror is a radio program that is published as a podcast. The content through traditional broadcast channels is "mirrored" in the world of podcasting, and, as such, the podcast content is no different from the broadcast content. While not as original, the Mirror format allows for full usage and exploitation of existing content, thereby populating the world of podcasting, and harnessing new technologies to maintain and grow one's audience. Both commercial and public radio employ the Mirror format quite extensively (the Sprout, in contrast, is when new, complementary content is created by the same producer, but that content resides only in the world of podcasting).

THE PODCAST ECOSPHERE

The Top 5 Podcast Platforms and Streaming Apps

Apple Podcasts, Google Podcasts, Spotify, Stitcher and TuneIn. These are the top directories of podcasts, with Apple responsible for the lion's share of content and activity. The changes in the world of podcasting service providers seem to increase in frequency every few weeks. In the last year alone, services like Breaker, Chartable and Spotify have either purchased, or been purchased by, other companies in the podcast ecosphere. So, maintain a constant monitor on who is doing what and which companies are being absorbed into the service offerings of other companies.

Apple Podcasts

Originally named iTunes, Apple Podcasts is a dominant player. A podcast not listed with Apple will not be discovered or listened to by many people. In addition to a major directory, Apple Podcasts is a listening service through an app as well through an iTunes account. A manual request by the podcaster is usually required for Apple to include a new podcast in its listing, a process that requires the RRS of the podcast and takes effect in a few as three to four days.

Google Podcasts

Google is one of those "we do everything" companies (they nearly do). Google is a place where a podcaster definitely needs to be listed. Another advantage of having a podcast listed with Google Podcasts is that program episodes will appear – and are playable – in Goggle's search results.

Spotify

Spotify is a company that seems to grow by the day. The Spotify app is ranked as the 2nd most popular destination for listeners in many countries. Spotify officially works with Anchor (since Anchor is owned by Spotify). In other words, anyone with an Anchor account, or anyone who uses Anchor to create and publish podcasts, is automatically pushed into Spotify's realm. And that realm is rapidly growing. Spotify is one of those companies that, mainly through strategic acquisitions, does many things.

Stitcher

Stitcher is a key podcast directory as well as a highly popular listening app. With eight million users and 65,000 podcasts, Stitcher also found additional value as one of the "hard-wired" apps that were standard features in earlier integrated cars. Stitcher's recent alliance with Midroll, an advertising network company, helps position the firm as a leader in podcasting.

TuneIn

Completing the Top Five, TuneIn is a directory and a listening app. Originally launched in 2002 as Radio Time, TuneIn features more than 100,000 radio stations from around the world (including AM, FM, HD, LP, digital and internet stations), and streams more than four million podcasts. The company partnered with Adobe Advertising Cloud in 2018 to integrate targeted audio ads to consumers via smart speakers and voice assistants. TuneIn also features audiobooks and live audio streaming of professional sports.

A podcaster resides in comfortable territory if he or she can confirm listing and availability through these five avenues.

HOW TO DISTRIBUTE YOUR PODCAST MORE WIDELY

Anchor represents an example of a simple and efficient way to make, publish, host and distribute a podcast. This company's recent alliance with Spotify demonstrates the value in using this service. While there are several directories and listening services that a podcaster needs to individually and manually approach for listing and inclusion, Anchor gets your podcast out there to a healthy number of places without any effort or outreach on the part of the podcaster.

Amazon Music, PlayerFM and IHeartRadio and three services to watch, each of which has grown substantially over the past year.

Buzzsprout is another great option that many podcasters use. This company helps podcasters create, host, promote and track their programs. They offer free-of-charge as well as highly affordable, high-quality services for those who want more customization. Buzzsprout can also facilitate getting a podcaster's content out there. Riverside.fm and Spreaker offer fee-based services for recording, hosting and monetizing, the quality of which many podcasters consider outstanding. Libsyn is another popular choice in a similar vein. RedCircle and Blubrry have moved decidedly into full-service models as well.

There are many more, too many, in fact, to list in this space; but all of which can be considered by a podcaster. Then there are those companies that will "catch" your podcast through its RSS, while others require a manual request for listing.

The key is getting your podcast out to the Top Five, and then working tirelessly to ensure inclusion and listing in as many more as possible – particularly wherever your audience listens to their podcasts. It's also important to remember that some listeners consume different types of podcasts from different places. A listener might use one app for certain types of content or genres, and use a different service for other genres or types of content. Confirming availability of your podcast in these (and there are more!) services and apps can also add convenience for your listeners and expanded outreach for your podcast: Bullhorn, PodBean, Playapod, Himalaya, Podcast Addict, Podcast Guru, Podchaser, RadioPublic, Audacy, Deezer, or Vurbl. YouTube is currently in its early stages of releasing a podcast feature for publishers, which could soon be another option of distribution.

In addition, there are language-specific or regionally-oriented podcast apps, such as iVoox (Spanish), Kuku FM (from India), Poddog (Japanese), Anghami (focusing on content in Arabic), Fyyd (based in Germany), or Pod LP (catering to the "next billion", that is economically-developing countries, offering audio content in multiple languages).

Audioburst is another useful option to consider. This is a company that built an AI engine that listens to the vast amount of audio content available, and analyzes and cuts the content into short clips to make it more discoverable on the internet. The company describes its services as an "Audio Search and Delivery Platform", which "makes audio content accessible to users, whenever they want it and wherever they are." The service automatically transcribes audio content, and the clips are easily shareable to Twitter, Facebook and other social media. The "burst" is the action of sharing the content chosen, something quite different from other podcast services. Submission is free, requiring an RSS feed.

OTHER USEFUL SERVICES IN DISTRIBUTION OUTREACH

Audio Audit

It is always useful for any content creator to confirm the quality of the product at the user's end. Podcasters can now check out a service that verifies the quality of the audio stream and recording. As the company's website states, "Audio Audit's automated QA can make your podcast look and sound more professional by improving your workflow. We highlight the areas of concern within an episode and show you how to fix them, improving overall quality."

Audio Audit checks for volume, loudness, silence, encoding, compression, metadata and more.

https://audioaudit.io/

Cast Feed Validator

This service is actually a Blubrry production, and is quite useful. Cast Feed Validator is a service that tests everything from artwork to server performance. Simply enter your podcast's RSS into the validator engine and your results will be generated.

The value in this service gives a nod to the fact that there are many podcast directories, and every podcaster should make the effort to ensure a broad and wide a distribution as possible of the podcast. Each directory needs to be able to read your RSS feed in order to get listed. As with any form of online or digital compatibility, everything also needs to be current, and the Cast Feed Validator helps achieve these goals.

https://www.castfeedvalidator.com/

STEVEN CHRISTIANSON

EDUCATIONAL USES FOR YOUR PODCAST

Beyond providing a forum for entertainment, information or narrative, podcasts have expanded utility in the fields of education/public messaging.

Schools worldwide use podcasting as instructional tools. A student using podcasts as educational tools enjoys the benefits of mobility without time constraints. For educators, simple and free podcast creation tools become invaluable aids in creating instruction and lessons. Increasingly, students use podcasts as a supplement to lectures and other course materials. Students rely on podcasts to enhance their understanding of lectures – which can help promote discussion and dialogue in the classroom setting. Podcasts can serve as excellent sources of review material prior to examinations in ways in which the audio reinforces the visual elements of the course materials. Absent or remote students have relied on podcasts, increasingly since 2020, to remain current in their required study materials. Podcasts provide an excellent source for recommended content. As an educational tool, podcasts can also be more accessible for students with mobility issues and other disabilities. In sum, whether relying solely on audio or incorporating video content into the podcast, podcasting in the world of education can increase motivation among students, help improve study habits, aid in exam preparation and outcomes, and enhance the mobility and accessibility of learning materials.

These educational and training benefits are not exclusive to the formal classrooms and world of students; they also apply to the training and orientation of volunteers and new staff, to tourists visiting local attractions, to residents receiving public service announcements, to annual staff training exercises, and to news and messaging updates from the executive offices of any type of organization.

RECENT DEVELOPMENTS

The ecosystem of podcasting is growing and evolving very fast. It is critical for the podcaster to keep abreast of the changes, and to maintain a keen eye towards the horizon of technology and acquisitions.

In 2021, Facebook announced an ambitious plan to offer podcasting, including creation tools, a hosting platform, listening options and much more. In less than a year, Facebook not only rebranded as Meta (with Facebook constituting one of the new company's main pillars), but everything to do with podcasting was abandoned with an announced date of termination.

More and more podcasters are doubling up as influencers. This trend feeds directly into an accompanying trend – podcasts and podcasters are becoming entrenched as adjuncts of marketing (or content creation and promotion). Brands increasingly turn to the world of podcasting for promotion. Another development sees a rise in live audio and streaming; social media platforms increasingly offer a place in their service for live audio (or video, as the case may be). And, of course, we're seeing significant changes in the number and names of players in the field, as companies expand, acquire or retract their partners and services.

Twitter has made announcements about moving slowly and cautiously into the world of podcasting. While issues related to corporate ownership and structure at Twitter are in possible transition, the possibility of podcasting on this platform is definitely something to watch. For example, "Twitter Create" and "Twitter Spaces" jointly function to allow podcasters enhanced ability for promotion, listener engagement and monetization.

Twitter's "Create" is a dedicated space in the Twitter universe that allows podcasters to share live audio content, to post photos in real time, and to create and share audiograms. "Spaces" allows hosts and guests to enjoy live audio conversations, easily accessible via captioning. A "space" can be pre-scheduled, notifying listeners in advance of how and when to receive the podcast. "Twitter Spaces" has attracted the newsrooms of organizations like NPR and *USA Today* for the features of "Spaces" that include live interviews, panel discussions, direct reports from the field, and even listener quizzes. Another of Twitter's podcasting features, "Tips", can link a podcaster's existing accounts (third-party payment services), such as Patreon or Bandcamp, allowing listeners to reward podcasters for their audio content.

YouTube is exploring its evolution into podcasting. YouTube is in the process of developing a new podcast discovery home within the app. Podcasters would be able to feed their audio shows via their RSS feeds, with YouTube offering audio ads, analytics, and more to help them showcase the content. Many podcasters already upload audio content to YouTube, and many Tubers also now have their own podcasts. The incorporation of podcasts into YouTube seems to recognize what podcasters and listeners already know: the service is about more than just video.

When *Podcasting Explained!* was released in 2021 Blubrry was one of several free podcasting streaming and listening services, whereas today Blubrry is a full fee-for-service podcast creation and development platform. As mentioned above, the company has also expanded into complementary products, such as the Cast Feed Validator.

Acast is expanding its reach, and purchased the previously independent service, RadioPublic, in February 2021.

Spotify continues to grow, particularly by means of absorbing smaller niche service companies, such as Chartable and the formally independent podcast streaming app, Breaker.

And video podcasting is slowly and tentatively carving out a niche for itself. In the spring of 2022, Spotify began offering limited "beta" versions of its new video podcasting feature.

So what was familiar and comfortable in the world of podcasting one year ago has altered in shape and form quite considerably.

What Is A Video Podcast?

A video podcast is simply a podcast with a video element.

That element could be static image or a full video recording of the studio, the podcast hosts and their guests.

How to Create a Video Podcast?

Each service provider or platform will provide details on file type, size and upload procedures. While there are variations, there are three main ways recommended here to create a video podcast.

1.   Static Image

By far the most basic, this method entails conversion of the audio file into a video file, and a single static image is featured as a background to the audio. Listeners see the static image where the video screen resides while the audio is played. Services like tunestotube.com can perform this conversion and upload to YouTube.

2. In-studio Video

This method entails the set-up of video equipment to record the production of the episode. At its most basic, this method requires a camera to record video. The video file typically requires editing, so video editing software is also required.

3. Remote Interview

Recording a remote interview happens when footage of multiple guests, situated somewhere else, are captured in video form. The result typically displays multiple "talking heads", as one might experience with online video meeting software. Software that captures such a video feed is needed. For example, Zoom captures and consolidates multiple webcams in one place. Riverside is also well-known for providing similar and very high quality software tools.

Transcripts

The need and demand for podcast transcripts have quickly evolved over the past year or so. No longer do transcripts represent a nice "addition" to the podcast.

Some jurisdictions are governed by accessibility laws which require transcription. Legal compliance aside, transcripts can be useful for listeners – for sharing and posting to social media – and for podcasters who then have the option to repurpose the audio content into such form as newsletters or email communication. Another benefit of transcription is that the podcast becomes "searchable" by word or phrase in a relatively simple way.

Google Recorder is a useful smartphone application that facilitates recorded audio and transcription. Another option is to use the

Google Docs Voice Typing tool. YouTube also offers transcripts. And increasingly, platforms, plug-ins and other service providers are including podcast transcription as part of the standard roster of features.

TERMINOLOGY IN THE WORLD OF PODCASTING

AAC

AAC are the initials for Advanced Audio Coding, which is a compressed format much like an MP3, with more efficiency and sound quality. AAC is also commonly used for storing audio files (music) on the internet, computers and portable music players.

Ambiance

This refers to the extraneous sounds while recording on location, such as ocean waves, wind or downtown traffic. Ambiance can be mixed into something that becomes a natural sound effect.

Aggregator

An aggregator program is used to collect and read RSS and Atom feeds. An aggregator may also be known as a newsreader, news aggregator or RSS aggregator. Like a podcatcher, aggregators "listen" for cues of new episodes. When an aggregator receives such a cue, user's content is updated, downloading a new episode.

Analytics

Podcast analytics refers to the measurement and analysis of all the data around podcasting. This data includes listener demographics, episode performance, and audience engagement.

AOI

An acronym used in this book, all-in-one (or AIO) is a production approach for podcast creation that is simple, relatively quick and seamless for the user. This approach is contrasted by Maximum Creative Control (see MCC in this glossary).

App

Abbreviated from "application", this refers to a computer program or software application designed to run on (typically) a mobile device.

Bit Depth

Bit depth determines the quality of your audio. It is a setting that can be adjusted prior to recording any audio. Bit depths typically are in 16, 24 and 32 options. Most podcasters should use a bit depth of 16.

Bit Rate

Bit Rate is measured in Kilobits per second (or Kbps). The bit rate tells you how many bits of data are in a one second of audio. The key takeaway with bit rates is - the higher the bit rate, the larger the file. A common standard for most podcasts (voice, background music, sound clips) is 128kbps.

Broadcast

This is the transmission of sound or pictures that are carried over distances using radio waves.

Condenser Microphone

A condenser microphone uses a phantom power source (or battery) to operate. Condenser mics are usually more sensitive and can pick up more of the audio in an environment. Also see this Glossary for Dynamic Microphone as a comparison.

CPM

CPM is an advertising rate, and stands for cost per mille or cost per 1,000 listeners.

Cross-Fade

Cross-fades create seamless transitions from one sound to another. One sound is fading out while another sound fades in.

Distribution

Distribution is when you share your podcast on listening apps like Spotify, Google Podcasts, and Apple Podcasts to reach a larger audience. RSS is a primary vehicle of sharing. See this glossary definition for RSS.

Download

Download means to receive data from a remote system over the internet.

Duck

This occurs when the volume of one track (say, the original language in an interview) is lowered so as to sound as if it plays in the background – one track "ducks" behind the other - while another audio file plays at a higher volume (say, the translated language).

Dynamic Microphone

Dynamic mics do not need an additional power supply to work. They tend to be less sensitive than condenser mics, which can reduce the risk of peaking or clipping audio. Also see this glossary for Condenser Microphone as a comparison.

Episode

This is a single installment of a podcast (or program, or show), a term somewhat analogous to a weekly episode of a traditional television program.

Flac

Flac is an acronym from Free Lossless Audio Codec. Flac is an audio compression algorithm which, audiophiles generally insist, is sonically superior to the more widely-known MP3 compression algorithm.

HAM Radio

Amateur radio, also known as HAM radio, is the use of a radio frequency spectrum for purposes of non-commercial exchange of messages, wireless experimentation, self-training, private recreation, radio-sport, contesting, and emergency communication.

ID3 Tags

ID3 is a metadata specification that allows information to be added to MP3 files. Information such as track title, artist, album and track number are placed within ID3 "tags" that identify the type of data contained within.

Lossless Audio

Audio is often compressed when converted to a digital file format. Compression helps reduce the size of the file. However, compression can lose some of the recording: frequencies at the highest and lowest ends can be lost. This results in "lossy audio". In contrast, "lossless" audio preserves all the frequencies, and occurs when a digital copy is identical to the original recording. Lossy formats include MP3 or AAC. Lossless formats include WAV files and FLAC.

Mirror

A podcast mirror, as referenced in this book, refers to content that originates from a radio broadcast and is duplicated in a podcast. Most radio stations have offer mirror content through podcasts, sometimes available through their websites, increasingly through dedicated distribution apps. The advent of the smartphone, and the mobility implications around app usage, significantly advanced this trend. Podcast mirrors are widely available.

Monetization

Monetization literally means to convert something into money. This occurs when something is turned into a revenue-generating activity, service, or asset. Podcast monetization can be direct or indirect. Direct podcast monetization is when the show is the thing you're selling. Profit is generated from the creation of original content, repurposing it, and granting exclusive access to paying members. Indirect podcast monetization refers to the use of a podcast as a tool to sell other things.

MP3

MP3 is a coding format for digital audio. Also known by its technical name, "MPEG Audio Layer-3" MP3 is a compressed audio file format, and was the dominant format to store music (audio) files on computers and portable music players.

MCC

Maximum Creative Control (as used in this book), in contrast to all-in-one (or AIO), is a production approach for podcast creation that allows and requires at least an introductory level of proficiency in recording and editing software. MCC allows greater creative input for the podcaster, but does require more time than AIO. See AIO in this glossary.

Performance

Performance refers to the measurement of how well (or poorly) podcast content is received and used by listeners. Primary measures of performance include: downloads and listeners; social media engagement; trackable calls to action; and rankings and reviews.

Pirate Radio

Pirate radio (or a pirate radio station) occurs when a radio operator broadcasts without a valid license. In some cases, radio stations are considered legal where the signal is transmitted, but illegal where the signals are received - especially when the signals cross a national boundary.

Platform (context of podcasts)

This is an online provider of content, typically entertainment (music, movies, etc.) that delivers the content via an internet connection to the subscriber's computer, TV or mobile device, and operates within its own infrastructure or digital ecosystem.

Podcast

A podcast is an episodic series of spoken word digital audio files made available on the internet that a user can download to a personal device for convenient listening. Streaming applications and podcasting services provide a convenient and integrated way to manage a personal consumption queue across many podcast sources and playback devices. New installments can be received by subscribers automatically. Podcasts are also referred to as programs or shows.

Podcatcher

A podcatcher is a piece of software designed to "catch" podcasts. Podcatchers allow for the collation and cataloging of podcasts, to be collected as new episodes are released, rather than individually seeking out podcasts to manually download and listen.

Public Domain

The public domain consists of all the creative work to which no exclusive intellectual property rights apply. Those rights may have expired, been forfeited, expressly waived, or may be inapplicable.

Royalty

A royalty is a payment made by one party to another that owns a particular asset, for the right to ongoing use of that asset.

Royalty-free

Royalty-free is a good thing for podcasters, and particularly relevant to the use of music and sound clips. This is material subject to copyright or other intellectual property rights that may be used without the need to pay royalties or license fees for each use.

RSS

Really Simple Syndication. RSS is a web feed that allows users and applications to access updates to websites in a standardized, computer-readable format. These feeds can, for example, allow a user to keep track of many different websites in a single aggregator.

Sample Rate

Sample rate refers to the measurement of how many data points of a sound wave are taken, analogous to the number of frames in a film (or frame rate). Generally speaking, the higher the sound rate, the better the reproduction of the sound is to the ear. Higher sample rates also result in larger file sizes.

Satellite Radio

Satellite radio is defined by the International Telecommunication Union (ITU)'s "ITU Radio Regulations" as a broadcasting-satellite service (waves are broadcast via satellite). It is available by subscription, mostly commercial-free, and offers subscribers more stations and a wider variety of programming options, formats and uncensored content than terrestrial radio.

Social media

Social media is a computer-based technology that facilitates the sharing of ideas, thoughts, and information through the building of virtual networks and communities. By design, social media is internet-based and gives users quick electronic communication of content. Social media can include websites and applications that enable users to create and share content or to participate in social networking.

Sprout

Podcast sprouts, as referenced in this book, are a more recent phenomenon. The sprout is an accompanying program, exclusively available through a radio station's podcast, which "sprouts" from daily content. Print media also use sprout podcasts to augment their print publications without duplicating the content (as in the case of mirrors). Sprouts can represent a highly creative way to maintain the terrestrial presence of the radio station, with new growth (or "sprouts") in the digital world.

Streaming

Streaming refers to a method of transmitting or receiving data (especially audio or video material) over a computer network as a steady, continuous flow, allowing playback to start while the rest of the data is still being received.

Terrestrial Radio

This is any radio signal that travels along the land that is broadcast from a land-based station and is received by land-based receivers.

Tuber

This is a content creator who uploads YouTube video or audio content, especially those with an audience base. Although the term can refer to anyone active on YouTube, it specifically means someone who creates content, and interacts with followers and others on YouTube via comments or other social networks.

Upload

An upload is the transfer of data from one computer to another, typically to one that is larger or remote from the user, or functioning as a server.

WAV

WAV is an audio file format, created by Microsoft, which has become a standard PC audio file format for everything from system and game sounds to CD-quality audio.

Webcast

A webcast is a media presentation (an event, such as a musical performance) which you can listen to or watch on the internet.

XLR

External Line Return. XLR-type microphones are the go-to standard for high-quality audio inputs, commonly used in professional audio equipment setups. The connectors are circular in design, and, many audiophiles insist, XLR mics send a more balanced signal that isolates noise.

FAQS ABOUT PODCASTING

I want to produce a podcast, but how do I get beyond my feelings of discomfort about my voice?

Discomfort over hearing one's own voice is quite common. In psychology, the phenomenon is called "voice confrontation", or "self-confrontation", which basically refers to not liking the sound of one's voice. For some screen actors, the discomfort never diminishes, which is why we often hear actors proclaiming, "I never watch my own work". How does one get beyond that point? The simple answer: practice, practice and time. There is no quick and easy answer, except to listen to your recorded speech over and over.

How do I name my podcast, and can I change the name later?

Do your best to create a name that is short and sweet. Try to think about the name when it is physically condensed, or visually represented, such as on a lapel button or a t-shirt. Some people use a play on words. For professional and business podcasts, one should always try to incorporate or reference the brand name. Think about the tone of your podcast. Consider a name in the context of how well or how easy it is to search. Avoid tongue-twisters. Say the name you have in mind repeatedly and aloud. Finally, run your ideas by a few other people in whom you confide.

Is there an ideal running time for each episode?

The short answer is NO, and it depends on the type of podcast content you are publishing. Scan the listings and consider the templates used by others in your genre or category. The variance is considerable: some podcasters, especially those with co-hosts or participants, publish opuses with running times that exceed an hour, sometimes more. Others that deliver commentary on current affairs can be snappier, typically averaging three to six minutes per episode. You will likely find your comfort zone once you get a clearer picture of who is listening, how and where they enjoy your podcast, as well as how many voices you include in an episode.

How often should I publish episodes?

Some podcasters publish weekly, others less frequently (every two weeks to three weeks), while some organization podcasts stick to a monthly schedule. That said, there are many podcasts that publish daily content. Reliability is the key factor in determining an appropriate publication frequency, both for listeners and for sponsors, as well as who your listeners and sponsors are.

Should I monetize my podcast?

Monetize when it is appropriate in your plan. One should never be overly concerned about monetizing until the podcast has an audience. After all, without listeners, there's nothing much to sponsor. One should generally have a minimum of 500-600 downloads per episode (some advisors might even double, or even triple, that number) before considering monetization. When monetization is appropriate for your podcast, consider working with an ad network, such as True Native Media (as one example).

Should I record in mono or stereo?

Some audiophiles insist that stereo provides a more balanced result and a nicer listening experience. Radio-styled dramas or podcasts with significant audio and musical effects may benefit from stereo. However, the typical conversational podcast will not receive any real enhancement with a stereo recording. Furthermore, mono recordings result in a smaller file size, and this can be a determining factor for many podcasters.

Do I need an intro and outro with theme music?

Ask yourself, "If my podcast were a TV show, what kind of theme song would it have?" How would you react to a program starting without any musical introduction? It is critical to have a good intro to your podcast. Although a potential listener might be interested in your topic, if you don't hook them in from the beginning, then you might not grab enough of their interest for them to continue listening – or, more importantly, encourage others to listen. Musical themes also help define and reinforce your brand: the listener associates the notes or the melody with you, your program, and the exciting content you share.

Is it necessary to add music or sounds in the body of the episode?

While not necessary, the use of music and sounds in the body of your narrative (the main part of the audio file for an episode) can guide the listener. Certain points in a commentary can be emphasized with the addition of sounds or music. Segues from one point, or one topic, to another are also suitable areas to apply audio effects. The key is to use such effects appropriately and sparingly. Some companies that specialize in podcast creation offer free libraries of sounds and music clips that can easily be inserted into an episode. Background music is also a great way to add texture and heighten listener interest. There are plenty of pay-per-use sites out there. Some, like FreeMusicArchive.org or DL Sounds, also offer royalty-free downloads in dozens of categories and genres.

Should I produce trailers and promos for my podcast?

It is always a good idea to produce a trailer for a podcast. Trailers can be distributed and shared through multiple communication means and social media. While some podcasters create trailers for individual episodes, most focus on a good trailer for the program. Once you have a trailer or promo, consider updating it cyclically (perhaps annually, perhaps more frequently).

How long does distribution take effect after publication?

While it depends on the platform, new episodes can appear within minutes. A 24-hour turnaround is usually safe to plan around. A brand new podcast (not just an episode) can take three to five days.

Why do streaming platforms ask "is this your podcast?" and "do you want to claim it?"

Claiming a podcast lets the creator see the show's analytics (for example, listening statistics and demographics). This type of information is useful to most podcasters. The process is simple: typically, after claiming, a company will send an email to the address associated with your podcast hosting account, providing instructions on how to verify and confirm the details of the show being claimed.

How do I let people know that I have a podcast?

Consider your inventory of lists and contacts. Build an email list (or outreach list) before your podcast launches. Make sure that you have some content ready for consumption. In other words, have three to five episodes ready for publication prior to launch. Consider your inventory of images, media articles, quotes and excerpts from your podcast that are shareable. Distribute your message to your contacts in your assembled inventory. Ask each recipient to listen to a selected episode, to identify the top two things he or she did not like about the episode, and to consider posting a review about the thing(s) he or she liked the most. There are multiple and assorted other strategies for letting people know about a podcast, but these tips can represent the simplest and most straight-forward.

Can I include guests who participate via telephone?

Absolutely. There are add-ons and apps through your smartphone that record and save audio files in different formats. Once you have the audio file (and expressed permission by the interviewee to use the content of the conversation), drop the audio it into your editing software, and you're ready to mix it.

Phone interviews are increasingly popular. They are more convenient for the guest, which increases the chances of getting that person to participate in your podcast. Furthermore, a conversation over the phone can elicit a more relaxed atmosphere than physically sitting in a studio or at an audio station. And the more relaxed the conversation, the more authentic the content.

How do I receive and respond to feedback from listeners?

You can receive feedback in a number of ways: during your podcast, on social media, in your newsletter. You can ask people to email you or send you an audio message via voicemail (which is built directly into some platforms).

Is it better to produce a studio-quality, professional podcast?

The "Cadillac Approach" is perfectly suited for some people. Certainly, the quality is nearly always undeniably higher. However, is it a "better" option? Not always, and sometimes the slickness of the production can reduce both the authenticity and interest of listeners. If people want to hear professional voices speaking through highly expensive studio and other technical equipment, they'll listen to the radio. Podcasts are an alternative. Aim for your best possible quality, but don't try to replicate what you might hear on your favourite radio station. In sum, be yourself (or the brand persona that your followers think you are) and not someone either you or an advisor suggests you be. Read more about the Cadillac Approach in Chapter 12 of *Podcasting – Explained!*

Why is it called a podcast?

The word is actually a combination, or something known as a *portmanteau* (similar to the blend of words that gives us "sitcom", "BREXIT", or "smog"). *Guardian* columnist and BBC journalist, Ben Hammersley, first coined the term by combining the words iPod and broadcast.

I want to do a podcast by recording my own audio. But that's all I want to do: talk into a microphone. Is there a way for me to get that recorded audio file into another person's hands so he or she can do all the editing, post-production, publishing and distributing?

Absolutely. Consider what some call "The Cadillac Approach", also the title of Chapter 12 in *Podcasting – Explained!*, which offers greater detail in how to approach such an option.

What if I don't have any particular focus for a topic, and I want to discuss several things that I'm interested in?

A podcaster enjoys one of the purest forms of audio freedom: the podcaster can talk about whatever sparks an interest or fuels a passion. That said, imagine if you were browsing an inventory of podcasts. What do you want to hear? More importantly, what would you likely subscribe to? Consider how someone catalogs your program: how does someone find you? Enjoy the freedom, but make sure that your program is searchable and easy to discover.

REVIEW SHEET 1:

A PRACTICAL PODCASTER'S SELF-ASSESSMENT

Every now and then, refer back to this Self-Assessment, and try to answer each question as honestly as possible.

Are you creating a podcast for your own personal reasons? Or are you doing it for professional reasons?

If it is personal, are you (a) sharing knowledge and experience, (b) hoping to inform or teach, (c) wanting to learn new skills, or (d) a combination of all or (e) none of the above?

If it is professional, have you been contracted to produce a podcast for someone else, or are you creating a podcast that is connected to your own business?

What do you want to share?

Do you want to lecture, rant, complain, inform or persuade?

Would you say your focus is informative and educational? Or is it more entertaining?

Do you want to talk "at" listeners or engage "with" them?

Who are you in this podcast, and is that persona any different from who you are privately?

Is there an expectation to make money from this podcast?

Are you a celebrity of some sort on some level (answer yes, even if you are identified in a Wikipedia page, listed in your local Who's Who, or if your name and face are on local billboards and advertisements)?

Is there any pertinent visual material that a listener would need to see?

Do you send out bulk emails, newsletters, briefings or other bulk communication?

Does your firm, or the firm you are working with, use an intranet?

Do you, or the company you're working with, have any remote or field staff?

Do you have a budget?

How often do you want to publish episodes?

How much time are you able to dedicate each week to producing, publishing and distributing your podcast?

Do you have an episode or two in mind?

How would you describe your podcast in not more than six words?

What is your inventory of online, traditional content and promotional properties? Do you have:

- An existing podcast
- A website
- Any social media accounts or pages
- A blog
- A newsletter

- A company and/or collegial intranet
- Radio ads
- TV ads
- Video material published on YouTube or any other streaming platform?
- Newspaper ads
- Billboards (including transit posters)
- Online banner ads
- Social media ad placement
- Direct mail

REVIEW SHEET 2:

SOFTWARE REQUIREMENTS

Whether you want simple-and-easy or maximum creative control, the budding podcaster requires an account with a podcast service, access to that service on one's home computer and mobile app, a digital audio workstation (necessary for most, although some podcasters choose to rely exclusively on their podcast service's recording and editing features), and, for those wanting enhanced creativity, legal access to a website that offers sound files.

Do you want the greatest possible amount of creative control or something simple and easy?

If simple-and-easy is your preferred option, then the app you choose will also be your centre of production. Such apps do much of the work for the podcaster, including assigning an RSS, some distribution, archiving, etc. Some even provide facility for basic recording and editing without the need for a digital audio workstation (DAW). Most podcasters, however, will also require a DAW, which receives the recorded file and offers features for editing, tagging and saving audio files to uploaded later into a podcast service.

Select your podcast service, and download the app on your synched devices. Suggested apps include (but are not limited to): PodBean, Buzzsprout, SquadCast, Anchor (by Spotify), Swellcast, Riverside.fm, Blubrry, and Alitu.

Familiarize yourself with the app, and explore other users' feedback as well as any YouTube demos.

If you desire greater creative input, then consider using one of the following software programs: GarageBand (you will already have this if you are a Mac user), Logic Pro, Adobe Audition (formally Cool Edit), Audacity, Hindenburg Journalist or WavePad.

These programs can be quite sophisticated. Familiarize yourself with your program of choice. There are multiple demo videos available on YouTube. These programs receive your recorded audio and allow you to edit, add, delete and save for upload to the podcast service (and accompanying app) that you choose. Unlike the easy-and-simple approach, the podcaster will record and produce a finished audio file separate from the podcast service provider. The audio file may then be uploaded to the podcast service provider, rather than recording and editing straight into the service provider's system.

Finally, find a site that offers legal downloads of music and sound effects. As an example, consider DL Sounds Original Audio, www.dl-sounds.com, which offers totally free as well as inexpensive tracks for download. Another popular option to consider is Free Archive Music, at freearchivemusic.org.

REVIEW SHEET 3:

HARDWARE REQUIREMENTS

For the podcaster who wants creative flexibility, you will need the following:

- Mic (preferably dynamic)
- Audio interface (with cables), ideally that facilitates at least two mic inputs
- Mic stand or arm
- Headphones
- Pop filter
- Shock mount

In this setup for basic hardware, the mic plugs into the interface, and the interface is then connected to your computer. The main reason for the interface is to maximize sound input quality while facilitating more than one mic – a setup that is ideal for podcasters who co-host their programs or welcome guests as part of an interview format.

If you want a simple-and-easy approach, everything in the list above – except the audio interface – will be required. In this option the podcaster simply plugs the mic into the computer, recording directly into the system's native recording software or into the podcaster's chosen DAW. Most podcasters who choose this option tend to select a condenser mic with a USB connection, but with only one mic, the options for co-hosts and guests become limited. This option is ideal for the "solo" podcaster.

If you want a super simple option, source out a podcasting kit at your local store or online. Amazon, for example, offers a plethora of different types of kits, including bare-bones basic to some that include an audio interface.

REVIEW SHEET 4:

ON-THE-GO

One of the great things about podcasting is the mobility of recording, producing and listening. With podcasting, everything can be accomplished On-The-Go.

Some platforms and services offer a complete suite of recording options and publishing features through their smartphone applications.

For those podcasters who seek maximum creative control (MCC), consider obtaining the following items of equipment, which can be used interchangeably.

- A track mix/interface
- Zoom Podtrack P4 Podcast Recorder
- 2 dynamic mics with XLR cables
- 2 table mic stands
- 2 sets of headphones
- Lavalier mics that can plug directly into a mobile phone
- One laptop with USB connections and DAW software

REVIEW SHEET 5:

STUDIO OR ENVIRONMENTAL REQUIREMENTS

Your DAW will offer multiple approaches to reducing noise and enhancing the sound of the recorded audio. But always remember: the better the ingredients going in, the better the result at the end. Try to maximize your environment prior to recording.

Try to avoid building your setup close to a window. While this is not always possible, windowpanes can reflect sound, sometimes creating echo and a "tinny" sounding environment.

Take a "sound inventory" of your immediate surroundings. This would include listening carefully for anything that could be picked up in the recording, sounds coming for such as: a furnace, a refrigerator, the fan in your computer, analog clocks or watches that feature an audible ticking sound, fans and heaters, even the weather conditions outside.

Turn off the ringer and buzz notification on your phone. Be aware of any planned fire drills in your apartment, condo or office building. Pets tend to love the sound of our voices, which is wonderful in most circumstances – except those when we want to record clean audio without the sound of your dog or cat communicating back to you.

Your mic should be positioned a few inches lower than your mouth, and in a spot that allows you to sit naturally and speak in a forward-projecting manner.

Try to keep the mic at least 14 inches away from nearby walls, cubicle dividers, etc.

Consider foam panels from a local music store or online. These can be as inexpensive as a few dollars each. Bathroom towels or pillows can also be placed around your setup to dampen echo and help achieve a more authentic studio sound.

REVIEW SHEET 6:

MONETIZING YOUR PODCAST

Is your podcast worth money to anyone? Before monetizing your podcast, consider the following.

Listeners want something interesting, entertaining, possibly provocative. Listeners will not follow your podcast if your prime motive is finding a way to generate revenue. This will show.

Craft your expertise, and embrace your passion, around the topic and issues you want to discuss.

Develop your audience base.

Build an archive of at least a dozen episodes prior to implementing any monetization strategies.

Listener downloads should be attractive to any potential sponsor or partner. Do you have 600, 700 or 800 downloads per episode? Don't even worry about monetization until your program resides in that level of consumption (keeping in mind that most sponsors like to see significantly more).

Have you defined your key metrics? What analytics and measures can you present to a potential sponsor?

What are the top 5 unique features of your listener analytics that might peak the attention of a sponsor?

Like any business or operating plan, consider how (or if) monetization might work during the developmental stages of your podcast.

Ask yourself: As a business owner or representative, why would you sponsor a podcast like yours? In the position of that business person, what risks do you see?

Have you consulted with an ad network?

How much revenue do you project on a monthly basis?

Have you identified the methods of monetization that are not appropriate for your podcast, and have you explained why?

What DON'T you like about your podcast?

REVIEW SHEET 7:

PREPARING YOUR INTERVIEW - A FEW LESSONS FROM THE FIELD

Preparation is critical when deciding to include interviews in your podcast. Always treat your guest's time as more important than yours.

In so doing, consider some of the following tips of advice from lifelong radio broadcast journalist, Mike O'Sullivan, Los Angeles Bureau Chief of the Voice of America. Mike immersed himself in audio technologies in his youth, having received a HAM radio license as a young boy, and has since spent decades interviewing personalities that include children's author Dr. Suess; Star Trek creator Gene Rodenberry, as well as Star Trek's Mr. Sulu, George Takei; Iwao (Tak) Takamoto, who created Scooby Doo; US presidential candidate Ralph Nader; Olympic athletes at the Great Wall of China during the 2008 Olympic Games; and Joe Barbera of the animation production studio, Hanna Barbera.

These tips assume a post-edit (or a "cleaning up") of the interview/conversation prior to publication, and are based on an email conversation with Mike in April 2021.

Ask your guest, 'what is the biggest misunderstanding about the issue being discussed?'.

Alternatively, near the close of the interview, ask if there are any important points that were not covered in your conversation.

Always try to get responses with emotion, and not just the facts. For example, ask the guest "how did you feel" about the topic or point you are discussing. While the question sometimes does not elicit much response from the guest, it can serve as a point of opening up other important areas of conversation.

Remember, while you might use the entire audio for the interview, it is more likely that you will choose select questions and answers. For example, if your guest has embarked on a boring recitation of facts and data, try to summarize what the person is saying, then confirm your summary by asking if it is correct. You'll either get a simple "yes" (which doesn't help much except to verify), or a response with an emphatic follow-up that evolves into a conversation that is more engaging, passionate and suitable for the listener of your episode.

Well-known personalities, or celebrities, have a natural draw simply because of the wider awareness of their names (or brands). But, as Mike points out, some of the most interesting and engaging interviews can be with regular people whose public name recognition is not at a high level: a family farmer in attendance at a rural campaign rally for a presidential candidate; a refugee worker in a migrant centre; a woman whose employment focuses on studying sharks in Hawaii; or a young orthopedic surgeon working out of a hospital in Haiti during the aftermath of the 2010 earthquake.

Whenever walking into an interview with any guest, regardless of their level of public or celebrity cache, ensure that you have done your research about the guest and the topic, that you have a list of questions (that can evolve organically during conversation), that you have mentally practiced and rehearsed your questions and interactions, and that you have tested your equipment and that it is ready to record.

REVIEW SHEET 8:

AN EPISODE TO-DO LIST

Formulate the Idea.

Conduct relevant research on the topic, guest, etc.

Conduct research into possible complementary programs or events that might correspond with the topic of the episode (eg, Halloween, Oscars, BAFTAs or Emmys as the events that complement a podcast about costume design in film, stage and television).

Draft tentative script.

Review, Read and Re-read.

Check your recording levels and gain in a test recording.

Record an approximately 30-second intro/trailer, selecting standard back music, and determining if it will be used as a stand-alone trailer/teaser published independently of the episode, or an introduction to your episode that follows from one the next.

Check your recording levels.

Record the body of the episode.

Edit by checking for tempo – pauses, ums, mouth clicks, background noise, peaking, compression, filter curves - and apply cleanup filters in your DAW.

Are there any special points you want to highlight for your listeners, perhaps a call to action? Record and place these files accordingly.

Add music and sound files where desired.

Insert sponsor-related materials (a customized pre-recorded ad inserted into the episode). Remember that the more integrated the ad is with the content, the more cohesive and non-invasive a listening experience it will be for the audience.

Are you certain that all copyright issues have been addressed?

Add your standard program theme with music before the body, and add the closing standard program theme with music at the end.

When saving your audio file, enter the Metatags tags: Program Name, Season, Episode name, Episode Number, Host, Date, Notes/Description.

Do you want any special episode-specific artwork?

If you made a separate trailer that promotes the episode, publication should precede the episode, but be coordinated for episode live date.

Post and announce the trailer to social media, using it to promote the episode.

Formal publication of the episode.

Also ensure simultaneous posting of episode to podcast blog site. Ensure the creation and upload of transcripts. Post a description of the episode with tags for the page.

Link to any personal or company webpages dedicated to the podcast.

Ensure manual uploads where needed, taking note of media such as YouTube or YouTube podcasts, Twitter's audio feed, global distribution services that specialize in regions or languages (for example, Ivoox, Anghami, or Gaana).

Use auto-promote (linked accounts) or manually announce that the episode is live, and promote on all social media.

Interact with audience as much as possible.

Does your promotion require separate or additional artwork?

Have you considered merchandise?

Even after you have published to a podcast website, ensure inclusion into all other and related personal and business/company websites, regular newsletters and outreach material.

Consider creative uses of social audio, such as Discord or Clubhouse, wherever appropriate.

Save the episode somewhere external to the website (a dedicated storage device) and save a separate back-up copy.

REVIEW SHEET 9:

DEFINING YOUR TARGET AUDIENCE

Sometimes there is a certain amount of trial and error involved in getting to know one's true target audience. The tastes and listening behaviour of people in different places can sometimes surprise the content creator. That said, we can try to get a gut feel for what types of people will enjoy, and become attracted to, our podcasts. Aside from this introspective process, the comparison of podcast platforms later in this *Companion Handbook* is useful in this regard as it illustrates the importance of analytics – the data about our listeners.

Do you instinctively feel that your listeners are local, regional, national or global? Are there specific regions or countries that you think will be part of your target audience?

What are the current trends around your topic? For example, does your topic appear in news feeds, or is it something more obscure and not necessarily news-of-the-day?

Will your podcast appeal to a specific gender or no gender at all?

What do you envision your listeners doing while listening to your podcast? Will they be driving, performing chores around the home, or enjoying a lunch break at work?

Will your topic be directed at – or exclude – certain age groups? For example, do you plan on speaking about something adult-oriented (reviewing alcohol brands, for example) and not suitable for minors?

Why would a potential listener dedicate part of the day to enjoying your podcast – especially when compared to the tens of thousands of existing podcasts? What will you offer to a listener, or a group of listeners, that you feel is unique?

What complementary media do you think your listeners consume, and how do they consume such media? It is useful to understand, for example, if your listeners are mobile-oriented or desktop-oriented; if your listeners receive news and other current affairs from conventional television, radio or social media platforms; which streaming apps and podcast platforms are more popular in the regions you are considering; or if your listeners prefer books and magazines in digital form or hard copy?

What would compel you, as a listener, to take the time to download or listen to your program?

Do you feel your target audience has a lot of spare time or are they high-performance professionals who can't afford to spend excessive time on any given activity?

PODCAST
HOST PLATFORMS

Anchor

https://anchor.fm/
Unlimited bandwidth
Free

Owned by Spotify, Anchor is among the simplest and most user-friendly platforms. They boast what many industry analysts argue is one the industry's most effective services for mobile users, allowing podcasters to create episodes entirely from their mobile phones.

Audioboom

https://audioboom.com/
10,000 downloads
$9.99 to quote-based

Audioboom is competitive for sponsorships and monetization. They facilitate the coordination of multiple podcasts from one place. They boast strong distribution partnerships and integration with the Top 5 streaming and listening apps. Audioboom also offers analytics from each partner. With embedded players for social media and websites, they also integrate Midroll ad insertion.

Blurry

https://blubrry.com/
Unlimited bandwidth
$12 to $20 month

Blubrry provides unlimited bandwidth for all plans. While storage options are not as strong as other platforms, Blubrry also provides a free integrated WordPress website. They also provide superior analytics on a pricing model.

Buzzsprout   https://www.buzzsprout.com/
250gb per month
Free to $24

Buzzsprout is great for new podcasters and seasoned veterans alike. Its interface, both for podcasters and listeners, is elegant and quite simple. Buzzsprout is known for its superior audio tools, helping turn a podcast into studio-quality. Buzzsprout allows podcasters to create a video highlight that is shareable on social media.

Captivate   https://captivate.fm/
12,000 to 150,000 streams per month
$19 to $99

One of Captivate's strengths is helping podcasters to grow their podcasts. Functionality that enables the insertion of a call to action is a unique feature. They offer very good analytics that follow advertising industry standards. Captivate also includes options for contracting with audio influencers.

Castos   https://castos.com/
Unlimited bandwidth
$190/year-$990/year

Castos is another of the more well-known and established platforms. In addition to competitive pricing and storage plans, they offer a Simple Podcasting WordPress plugin as well as a customizable player for a client's website.

Libsyn   https://libsyn.com/
                                    Unlimited bandwidth
                                    $5-Quote-based

Libsyn is among the more well-known and established platforms. Their enterprise model, the pricing of which is quote-based, is particularly suited to larger or more publicly-known clients, such as government or high-end celebrity profiles.

Podbean   https://www.podbean.com/
                                    100GB per month on free plan
                                    Free - $99

Podbean is highly competitive in its paid plans. Their services include a podcast website, built-in Patreon support, and capacity for video podcasting.

RedCircle   https://www.redcircle.com/
                                    Unlimited bandwidth and storage
                                    Free to Quote-based

RedCircle offers robust analytics, a healthy spectrum of distribution that includes YouTube, and a noteworthy edge in monetization tools including specific features for subscription-based podcasts.

Simplecast   https://simplecast.com/
                                    20K-150K downloads per month
                                    $15-$88

Simplecast is highly user-friendly. The platform offers custom web domains.

Soundcloud   https://soundcloud.com/
256kbps
$15/month

SoundCloud excels in listener and fan engagement. They offer the ability to schedule episodes, with multiple tools available to engaging listeners with direct messages, and further interact through the platform's comments section.

Spreaker   https://www.spreaker.com/
Unlimited bandwidth
$7-$50

Spreaker's interface is very clean and user-friendly. Reaching listeners is a one-click operation. Podcasters can also initiate "live streams".

Transitor   https://transitor.fm/
10,000 - 150,000 downloads/streams per month
$15.83 - $82.50 per month

The big strength of Transitor is its grasp on handling private podcasts as well as providing excellent analytics.

PODCAST DIRECTORIES

Apple Podcasts, Google Podcasts, Spotify, TuneIn, Stitcher and Amazon Music constitute those places where your podcast definitely needs to be. Some hosts and platforms distribute to these services on your behalf, while some manual configuration or outreach may be necessary for a few.

In addition to these services, which comprise the lion's share or podcast streaming, there are many more, each valuable in its own right. Get your podcast on as many directories and services as possible. Visit the websites to check each service for availability of your podcast. If your podcast is not there, reach out with a request. In many cases, there will be an assigned area for podcasters to submit their RSS feed and to enter other identifying information, while others facilitate a podcaster's request for inclusion in the company's directory.

Acast

https://play.acast.com/

Add My Podcast

https://addmypodcast.com/

Anytime Podcast Player

https://anytimeplayer.app

Anghami

https://www.anghami.com/

Audacy

https://www.audacy.com/

Audioburst

https://search.audioburst.com/

BlackPodcasting

https://blackpodcasting.com/

Blubrry

https://blubrry.com/

Bullhorn

https://www.bullhorn.fm/

Castbox

https://castbox.fm/

Deezer

https://www.deezer.com/

Digital Podcast

https://www.digitalpodcast.com/

Fyyd

https://fyyd.de/

Fountain

https://foundtain.fm

Gaana

https://gaana.com/

Gpodder

https://gpodder.net

Himalaya

https://www.himalaya.com/

IHeartRadio

https://www.iheart.com/

Ivoox

https://www.ivoox.com/

JioSaavn

https://www.jiosaavn.com/

KukuFm

https://kalakaar.kukufm.com/

Lava

https://www.joinlava.com/

Listen Notes

https://www.listennotes.com/

Overcast

https://overcast.fm/

PocketCasts

https://www.pocketcasts.com/

PodBean

https://www.podbean.com/

PodcastAddict

https://podcastaddict.com/

PodFriend

https://web.podfriend.com/

Podcast Guru

https://podcastguru.io/

Podcast Index

https://podcastindex.org/

Podbay

https://podbay.fm/

Podchaser

https://www.podchaser.com/

Poddog

https://www.poddog.jp/

Podhero

https://podhero.com/

Podimo

https://podimo.com/

Podknife

https://podknife.com/

PodLP

https://podlp.com/

Podtail

https://podtail.com/de/

Podvine

https://podvine.com

Plex

https://app.plex.tv/

PlayerFM

https://player.fm/

RadioPublic

https://radiopublic.com/

Vurbl

https://vurbl.com/

SOURCES OF SOUND EFFECTS AND SOUND FILES

BBC Sound Effects

The BBC Sound Effects Archive, of the British Broadcasting Corporation, is available for personal, educational or research purposes. Be sure to check out the Terms of Use. There are over 33,000 clips from across the world from the past 100 years.

https://sound-effects.bbcrewind.co.uk/

DL Sounds Original Audio

Based in the Netherlands, this is an excellent source of free and fee-based sound files, loops and samples. The sounds are all original and exclusive to the site. Check out their FAQ to learn more about their free and royalty-free music, as well as their License Agreement.

https://www.dl-sounds.com/royalty-free/

Audio Jungle

This site is a marketplace for creators and buyers. They offer royalty-free music and audio tracks starting at $1 USD. The site boasts more than 1,900,000 tracks and sounds from their community of musicians and sound engineers.

https://audiojungle.net/

Free Music Archive

This archive is known for free-to-download music, from classical and urban to hard rock and blues, which is licensed under Creative Commons. Be sure to read their License Guide. They also link to a "pro" version, called Tribe of Noise (with licenses that start at €45).

https://freemusicarchive.org/

Freesound

Freesound is a sizable collaborative database of audio snippets, samples, and recordings, released under Creative Commons licenses that allow their reuse. Freesound is also a community that facilitates interaction among sound artists.

https://freesound.org/

Free Sound Effects

This site boasts more than 10,000 free sound effects available for personal and educational use, and over 100,000 sound effects that come with a license for commercial use. For podcasters who require unlimited use in their projects, Free Sound Effects offers a lifetime license for a one-time fee. Be sure to read their License Agreement.

https://www.freesoundeffects.com/

Zapsplat

With more than 106,000 sound files, Zapsplat is a fast-growing free sound effects and royalty-free music library with file formats available in MP3 or WAV. Their files can be used in commercial projects, including broadcast, TV, film, radio, games, podcasts, and advertising. Their sounds are 100% royalty-free and safe for use in commercial and non-commercial applications. Read their Licensing Agreement.

https://www.zapsplat.com/

PRODUCTION PLANNER

Tip: *Consider checking out podnews.net or any of the podcast-focused subreddits. These forums frequently offer a lot of great advice, tips and interesting information.*

Podcast Name: _____

Recording Date: _____

Publication Date: _____

Notes of Relevance: _____

_____

_____

Host(s): _____

Guest(s): _____

Episode Number: _____ **Date:** _____

Episode Name: _____

Season Number: _____

Season Name: _____

RSS and podcast host/service provider: _____

_____

DAW: _____

Running Time: _____

Manual Uploads: _____

_____

_____

Categories: _____

Social Media Posts: _____

_____

Background Music: _____

Audio Clips: _____

Sponsorship and Placement: _____

_____

Metadata, ID Tags: _____

Artwork Used: _____

Other Notes: _____

_____

_____

Podcast Name: _____

Recording Date: _____

Publication Date: _____

Notes of Relevance: _____

_____

_____

Host(s): _____

Guest(s): _____

Episode Number: _____ Date: _____

Episode Name: _____

Season Number: _____

Season Name: _____

RSS and podcast host/service provider: _____

_____

DAW: _____

Running Time: _____

Manual Uploads: _____

_____

_____

Categories: _____

Social Media Posts: _____

_____

Background Music: _____

Audio Clips: _____

Sponsorship and Placement: _____

_____

Metadata, ID Tags: _____

Artwork Used: _____

Other Notes: _____

_____

_____

_____

Tip: *Research shows that users (or consumers prefer to receive promo content through their email to a far greater degree that through any other form of communication. Email is an effective way to encourage listeners to engage.*

Podcast Name: _____

Recording Date: _____

Publication Date: _____

Notes of Relevance: _____

_____

_____

Host(s): _____

Guest(s): _____

Episode Number: _____ **Date:** _____

Episode Name: _____

Season Number: _____

Season Name: _____

RSS and podcast host/service provider: _____

_____

DAW: _____

Running Time: _____

Manual Uploads: _____

_____

_____

Categories: _____

Social Media Posts: _____

_____

Background Music: _____

Audio Clips: _____

Sponsorship and Placement: _____

_____

Metadata, ID Tags: _____

Artwork Used: _____

Other Notes: _____

_____

_____

Podcast Name: _____

Recording Date: _____

Publication Date: _____

Notes of Relevance: _____

_____

_____

Host(s): _____

Guest(s): _____

Episode Number: _____   **Date:** _____

Episode Name: _____

Season Number: _____

Season Name: _____

RSS and podcast host/service provider: _____

_____

DAW: _____

Running Time: _____

Manual Uploads: _____

_____

_____

Categories: _____

Social Media Posts: _____

_____

Background Music: _____

Audio Clips: _____

Sponsorship and Placement: _____

_____

Metadata, ID Tags: _____

Artwork Used: _____

Other Notes: _____

_____

_____

Tip: *View the podcast as a source of information, as well as content that can feed directly into your organization's existing inventory of promotional and outreach tools, and vice versa.*

Podcast Name: _____

Recording Date: _____

Publication Date: _____

Notes of Relevance: _____

_____

_____

Host(s): _____

Guest(s): _____

Episode Number: _____ Date: _____

Episode Name: _____

Season Number: _____

Season Name: _____

RSS and podcast host/service provider: _____

_____

DAW: _____

Running Time: _____

Manual Uploads: _____

_____

_____

Categories: _____

Social Media Posts: _____

_____

Background Music: _____

Audio Clips: _____

Sponsorship and Placement: _____

_____

Metadata, ID Tags: _____

Artwork Used: _____

Other Notes: _____

_____

_____

Tip: *The more the podcaster understands his or her audience, and how that audience accesses and interacts with the podcast, the greater the potential for monetization (or any type of growth and evolution).*

Podcast Name: _____

Recording Date: _____

Publication Date: _____

Notes of Relevance: _____

_____

_____

Host(s): _____

Guest(s): _____

Episode Number: _____ Date: _____

Episode Name: _____

Season Number: _____

Season Name: _____

RSS and podcast host/service provider: _____

_____

DAW: _____

Running Time: _____

Manual Uploads: _____

_____

_____

Categories: _____

Social Media Posts: _____

_____

Background Music: _____

Audio Clips: _____

Sponsorship and Placement: _____

_____

Metadata, ID Tags: _____

Artwork Used: _____

Other Notes: _____

_____

_____

Podcast Name: _____

Recording Date: _____

Publication Date: _____

Notes of Relevance: _____

_____

_____

Host(s): _____

Guest(s): _____

Episode Number: _____   Date: _____

Episode Name: _____

Season Number: _____

Season Name: _____

RSS and podcast host/service provider: _____

_____

DAW: _____

Running Time: _____

Manual Uploads: _____

_____

_____

Categories: _____

Social Media Posts: _____

_____

Background Music: _____

Audio Clips: _____

Sponsorship and Placement: _____

_____

Metadata, ID Tags: _____

Artwork Used: _____

Other Notes: _____

_____

_____

Tip: *Continually share both the existence of the podcast as well as the individual episodes. The act of sharing is fundamental to the successful promotion of content.*

Podcast Name: _____

Recording Date: _____

Publication Date: _____

Notes of Relevance: _____

_____

_____

Host(s): _____

Guest(s): _____

Episode Number: _____   Date: _____

Episode Name: _____

Season Number: _____

Season Name: _____

RSS and podcast host/service provider: _____

_____

DAW: _____

Running Time: _____

Manual Uploads: _____

_____

_____

Categories: _____

Social Media Posts: _____

_____

Background Music: _____

Audio Clips: _____

Sponsorship and Placement: _____

_____

Metadata, ID Tags: _____

Artwork Used: _____

Other Notes: _____

_____

_____

Podcast Name: _____

Recording Date: _____

Publication Date: _____

Notes of Relevance: _____

_____

_____

Host(s): _____

Guest(s): _____

Episode Number: _____   Date: _____

Episode Name: _____

Season Number: _____

Season Name: _____

RSS and podcast host/service provider: _____

_____

DAW: _____

Running Time: _____

Manual Uploads: _____

_____

_____

Categories: _____

Social Media Posts: _____

_____

Background Music: _____

Audio Clips: _____

Sponsorship and Placement: _____

_____

Metadata, ID Tags: _____

Artwork Used: _____

Other Notes: _____

_____

_____

Tip: *Get to know where you do well with your episodes - and where you don't.*

Podcast Name: _____

Recording Date: _____

Publication Date: _____

Notes of Relevance: _____

_____

_____

Host(s): _____

Guest(s): _____

Episode Number: _____ Date: _____

Episode Name: _____

Season Number: _____

Season Name: _____

RSS and podcast host/service provider: _____

_____

DAW: _____

Running Time: _____

Manual Uploads: _____

_____

Categories: _____

Social Media Posts: _____

_____

Background Music: _____

Audio Clips: _____

Sponsorship and Placement: _____

_____

Metadata, ID Tags: _____

Artwork Used: _____

Other Notes: _____

_____

_____

Podcast Name: _____

Recording Date: _____

Publication Date: _____

Notes of Relevance: _____

_____

_____

Host(s): _____

Guest(s): _____

Episode Number: _____ **Date:** _____

Episode Name: _____

Season Number: _____

Season Name: _____

RSS and podcast host/service provider: _____

_____

DAW: _____

Running Time: _____

Manual Uploads: _____

_____

_____

Categories: _____

Social Media Posts: _____

_____

Background Music: _____

Audio Clips: _____

Sponsorship and Placement: _____

_____

Metadata, ID Tags: _____

Artwork Used: _____

Other Notes: _____

_____

_____

Tip: *While you may start out by coordinating your distribution and outreach on multiple social media channels simultaneously, you'll likely to do better on some more than others. Maintain a constant monitor on analytics every time you communicate something.*

Podcast Name: _____

Recording Date: _____

Publication Date: _____

Notes of Relevance: _____

_____

_____

Host(s): _____

Guest(s): _____

Episode Number: _____ **Date:** _____

Episode Name: _____

Season Number: _____

Season Name: _____

RSS and podcast host/service provider: _____

_____

DAW: _____

Running Time: _____

Manual Uploads: _____

_____

_____

Categories: _____

Social Media Posts: _____

_____

Background Music: _____

Audio Clips: _____

Sponsorship and Placement: _____

_____

Metadata, ID Tags: _____

Artwork Used: _____

Other Notes: _____

_____

_____

_____

Tip: *Launch with an inventory. In launching a podcast, it is best to have a minimum of three episodes complete and ready for immediate publication.*

Podcast Name: _____

Recording Date: _____

Publication Date: _____

Notes of Relevance: _____

_____

_____

Host(s): _____

Guest(s): _____

Episode Number: _____ Date: _____

Episode Name: _____

Season Number: _____

Season Name: _____

RSS and podcast host/service provider: _____

_____

DAW: _____

Running Time: _____

Manual Uploads: _____

_____

_____

Categories: _____

Social Media Posts: _____

_____

Background Music: _____

Audio Clips: _____

Sponsorship and Placement: _____

_____

Metadata, ID Tags: _____

Artwork Used: _____

Other Notes: _____

_____

_____

Podcast Name: _____

Recording Date: _____

Publication Date: _____

Notes of Relevance: _____

_____

_____

Host(s): _____

Guest(s): _____

Episode Number: _____   Date: _____

Episode Name: _____

Season Number: _____

Season Name: _____

RSS and podcast host/service provider: _____

_____

DAW: _____

Running Time: _____

Manual Uploads: _____

_____

_____

Categories: _____

Social Media Posts: _____

_____

Background Music: _____

Audio Clips: _____

Sponsorship and Placement: _____

_____

Metadata, ID Tags: _____

Artwork Used: _____

Other Notes: _____

_____

_____

Tip: *Does your recorded content have a lot of mouth-clicks? There are plug-ins available for the DAW which help identify and eliminate mouth-clicks. Other experienced professionals in the world of voice audio suggest taking a bite or two of an apple prior to recording. Bottom line: mouth-clicks happen with everybody, and there is no single solution.*

Podcast Name: _____

Recording Date: _____

Publication Date: _____

Notes of Relevance: _____

_____

_____

Host(s): _____

Guest(s): _____

Episode Number: _____   Date: _____

Episode Name: _____

Season Number: _____

Season Name: _____

RSS and podcast host/service provider: _____

_____

DAW: _____

Running Time: _____

Manual Uploads: _____

_____

_____

Categories: _____

Social Media Posts: _____

_____

Background Music: _____

Audio Clips: _____

Sponsorship and Placement: _____

_____

Metadata, ID Tags: _____

Artwork Used: _____

Other Notes: _____

_____

_____

Tip: *Whatever program format you choose, try to stick with it. Consistency is critical for listeners. Once you land on your preferred length of episode, style and format of presentation, or publication frequency, maintain consistency.*

Podcast Name: _____

Recording Date: _____

Publication Date: _____

Notes of Relevance: _____

_____

_____

Host(s): _____

Guest(s): _____

Episode Number: _____ Date: _____

Episode Name: _____

Season Number: _____

Season Name: _____

RSS and podcast host/service provider: _____

_____

DAW: _____

Running Time: _____

Manual Uploads: _____

_____

_____

Categories: _____

Social Media Posts: _____

_____

Background Music: _____

Audio Clips: _____

Sponsorship and Placement: _____

_____

Metadata, ID Tags: _____

Artwork Used: _____

Other Notes: _____

_____

_____

Tip: *Optimize Social Media. Get it Everywhere. Announce the launch of the podcast itself. Announce each new episode. This activity must now be ongoing and routine for you as a podcaster.*

Podcast Name: _____

Recording Date: _____

Publication Date: _____

Notes of Relevance: _____

_____

_____

Host(s): _____

Guest(s): _____

Episode Number: _____ **Date:** _____

Episode Name: _____

Season Number: _____

Season Name: _____

RSS and podcast host/service provider: _____

_____

DAW: _____

Running Time: _____

Manual Uploads: _____

_____

_____

Categories: _____

Social Media Posts: _____

_____

Background Music: _____

Audio Clips: _____

Sponsorship and Placement: _____

_____

Metadata, ID Tags: _____

Artwork Used: _____

Other Notes: _____

_____

_____

Tip: *Did you know that you can even create a simple podcast and individual episodes from your smartphone? From recording, to basic editing, and adding sound effects or background music, all of this can be accomplished on-the-go through the use of a smartphone and certain apps.*

Podcast Name: _____

Recording Date: _____

Publication Date: _____

Notes of Relevance: _____

_____

_____

Host(s): _____

Guest(s): _____

Episode Number: _____   Date: _____

Episode Name: _____

Season Number: _____

Season Name: _____

RSS and podcast host/service provider: _____

_____

DAW: _____

Running Time: _____

Manual Uploads: _____

_____

_____

Categories: _____

Social Media Posts: _____

_____

Background Music: _____

Audio Clips: _____

Sponsorship and Placement: _____

_____

Metadata, ID Tags: _____

Artwork Used: _____

Other Notes: _____

_____

_____

Podcast Name: _____

Recording Date: _____

Publication Date: _____

Notes of Relevance: _____

_____

_____

Host(s): _____

Guest(s): _____

Episode Number: _____   **Date:** _____

Episode Name: _____

Season Number: _____

Season Name: _____

RSS and podcast host/service provider: _____

_____

DAW: _____

Running Time: _____

Manual Uploads: _____

_____

_____

Categories: _____

Social Media Posts: _____

_____

Background Music: _____

Audio Clips: _____

Sponsorship and Placement: _____

_____

Metadata, ID Tags: _____

Artwork Used: _____

Other Notes: _____

_____

_____

Tip: *If you reference other brands, use music or sounds that may be owned by someone else, do your homework and confirm their status, possibly even reaching out to the registered owner for permission.*

Podcast Name: _____

Recording Date: _____

Publication Date: _____

Notes of Relevance: _____

_____

_____

Host(s): _____

Guest(s): _____

Episode Number: _____ Date: _____

Episode Name: _____

Season Number: _____

Season Name: _____

RSS and podcast host/service provider: _____

_____

DAW: _____

Running Time: _____

Manual Uploads: _____

_____

_____

Categories: _____

Social Media Posts: _____

_____

Background Music: _____

Audio Clips: _____

Sponsorship and Placement: _____

_____

Metadata, ID Tags: _____

Artwork Used: _____

Other Notes: _____

_____

_____

Podcast Name:

Recording Date:

Publication Date:

Notes of Relevance:

Host(s):

Guest(s):

Episode Number: **Date:**

Episode Name:

Season Number:

Season Name:

RSS and podcast host/service provider: _____

_____

DAW: _____

Running Time: _____

Manual Uploads: _____

_____

_____

Categories: _____

Social Media Posts: _____

_____

Background Music: _____

Audio Clips: _____

Sponsorship and Placement: _____

_____

Metadata, ID Tags: _____

Artwork Used: _____

Other Notes: _____

_____

_____

Tip: *Whenever possible, turn off anything in the vicinity that creates noise (phones, fans, home appliances).*

Podcast Name: _____

Recording Date: _____

Publication Date: _____

Notes of Relevance: _____

_____

_____

Host(s): _____

Guest(s): _____

Episode Number: _____ Date: _____

Episode Name: _____

Season Number: _____

Season Name: _____

RSS and podcast host/service provider: _____

_____

DAW: _____

Running Time: _____

Manual Uploads: _____

_____

_____

Categories: _____

Social Media Posts: _____

_____

Background Music: _____

Audio Clips: _____

Sponsorship and Placement: _____

_____

Metadata, ID Tags: _____

Artwork Used: _____

Other Notes: _____

_____

_____

Tip: *Most hard surfaces create reverb; most soft surfaces absorb sound.*

Podcast Name: _____

Recording Date: _____

Publication Date: _____

Notes of Relevance: _____

_____

_____

Host(s): _____

Guest(s): _____

Episode Number: _____ Date: _____

Episode Name: _____

Season Number: _____

Season Name: _____

RSS and podcast host/service provider: _____

_____

DAW: _____

Running Time: _____

Manual Uploads: _____

_____

_____

Categories: _____

Social Media Posts: _____

_____

Background Music: _____

Audio Clips: _____

Sponsorship and Placement: _____

_____

Metadata, ID Tags: _____

Artwork Used: _____

Other Notes: _____

_____

_____

Tip: *Whatever your choice of microphone, choose one ideal for podcasting – one that picks up the immediate sound in its front, and omits sounds behind or beside.*

Podcast Name: _____

Recording Date: _____

Publication Date: _____

Notes of Relevance: _____

_____

_____

Host(s): _____

Guest(s): _____

Episode Number: _____ **Date:** _____

Episode Name: _____

Season Number: _____

Season Name: _____

RSS and podcast host/service provider: _____

_____

DAW: _____

Running Time: _____

Manual Uploads: _____

_____

_____

Categories: _____

Social Media Posts: _____

_____

Background Music: _____

Audio Clips: _____

Sponsorship and Placement: _____

_____

Metadata, ID Tags: _____

Artwork Used: _____

Other Notes: _____

_____

_____

Tip: *Sound dampeners can help give your recording that "studio" quality. A bathroom towel often does the trick. Towels or pillows can significantly dampen sound and echo.*

Podcast Name: _____

Recording Date: _____

Publication Date: _____

Notes of Relevance: _____

_____

_____

Host(s): _____

Guest(s): _____

Episode Number: _____ Date: _____

Episode Name: _____

Season Number: _____

Season Name: _____

RSS and podcast host/service provider: _____

_____

DAW: _____

Running Time: _____

Manual Uploads: _____

_____

_____

Categories: _____

Social Media Posts: _____

_____

Background Music: _____

Audio Clips: _____

Sponsorship and Placement: _____

_____

Metadata, ID Tags: _____

Artwork Used: _____

Other Notes: _____

_____

_____

Tip: *The pop filter reduces the "puff" sounds we make when pronouncing certain consonants, like "p" or "b" (known as "plosives").*

Podcast Name: _____

Recording Date: _____

Publication Date: _____

Notes of Relevance: _____

_____

_____

Host(s): _____

Guest(s): _____

Episode Number: _____   **Date:** _____

Episode Name: _____

Season Number: _____

Season Name: _____

RSS and podcast host/service provider: _____

_____

DAW: _____

Running Time: _____

Manual Uploads: _____

_____

_____

Categories: _____

Social Media Posts: _____

_____

Background Music: _____

Audio Clips: _____

Sponsorship and Placement: _____

_____

Metadata, ID Tags: _____

Artwork Used: _____

Other Notes: _____

_____

_____

_____

Tip: *The Reddit subs, r/podcasting or r/podcast, offer the newcomer a wealth of opinions and experiences from those in the community, useful for beginners and seasoned experts alike. YouTube also features a wealth of content, from introductory lessons to video podcasting.*

Podcast Name:

Recording Date:

Publication Date:

Notes of Relevance:

Host(s):

Guest(s):

Episode Number: **Date:**

Episode Name:

Season Number:

Season Name:

RSS and podcast host/service provider: _____

_____

DAW: _____

Running Time: _____

Manual Uploads: _____

_____

_____

Categories: _____

Social Media Posts: _____

_____

Background Music: _____

Audio Clips: _____

Sponsorship and Placement: _____

_____

Metadata, ID Tags: _____

Artwork Used: _____

Other Notes: _____

_____

_____

Podcast Name: _____

Recording Date: _____

Publication Date: _____

Notes of Relevance: _____

_____

_____

Host(s): _____

Guest(s): _____

Episode Number: _____ Date: _____

Episode Name: _____

Season Number: _____

Season Name: _____

RSS and podcast host/service provider: _____

_____

DAW: _____

Running Time: _____

Manual Uploads: _____

_____

Categories: _____

Social Media Posts: _____

_____

Background Music: _____

Audio Clips: _____

Sponsorship and Placement: _____

_____

Metadata, ID Tags: _____

Artwork Used: _____

Other Notes: _____

_____

_____

Podcast Name: _____

Recording Date: _____

Publication Date: _____

Notes of Relevance: _____

_____

_____

Host(s): _____

Guest(s): _____

Episode Number: _____   **Date:** _____

Episode Name: _____

Season Number: _____

Season Name: _____

RSS and podcast host/service provider: _____

_____

DAW: _____

Running Time: _____

Manual Uploads: _____

_____

_____

Categories: _____

Social Media Posts: _____

_____

Background Music: _____

Audio Clips: _____

Sponsorship and Placement: _____

_____

Metadata, ID Tags: _____

Artwork Used: _____

Other Notes: _____

_____

_____

Tip: *Routinely visit Reddit subs and podnews.net for tips, updates, news and other happenings in the world of audio media and podcasting.*

Podcast Name: _____

Recording Date: _____

Publication Date: _____

Notes of Relevance: _____

_____

_____

Host(s): _____

Guest(s): _____

Episode Number: _____ Date: _____

Episode Name: _____

Season Number: _____

Season Name: _____

RSS and podcast host/service provider: _____

_____

DAW: _____

Running Time: _____

Manual Uploads: _____

_____

_____

Categories: _____

Social Media Posts: _____

_____

Background Music: _____

Audio Clips: _____

Sponsorship and Placement: _____

_____

Metadata, ID Tags: _____

Artwork Used: _____

Other Notes: _____

_____

_____

Podcast Name: _____

Recording Date: _____

Publication Date: _____

Notes of Relevance: _____

_____

_____

Host(s): _____

Guest(s): _____

Episode Number: _____   Date: _____

Episode Name: _____

Season Number: _____

Season Name: _____

RSS and podcast host/service provider: _____

_____

DAW: _____

Running Time: _____

Manual Uploads: _____

_____

_____

Categories: _____

Social Media Posts: _____

_____

Background Music: _____

Audio Clips: _____

Sponsorship and Placement: _____

_____

Metadata, ID Tags: _____

Artwork Used: _____

Other Notes: _____

_____

_____

Tip: *When you error in recording, snap your finger and pause. You'll see the error as a sound spike and space, which helps you identify the error and edit it out later.*

Podcast Name: _____

Recording Date: _____

Publication Date: _____

Notes of Relevance: _____

_____

_____

Host(s): _____

Guest(s): _____

Episode Number: _____ **Date:** _____

Episode Name: _____

Season Number: _____

Season Name: _____

RSS and podcast host/service provider: _____

_____

DAW: _____

Running Time: _____

Manual Uploads: _____

_____

Categories: _____

Social Media Posts: _____

_____

Background Music: _____

Audio Clips: _____

Sponsorship and Placement: _____

_____

Metadata, ID Tags: _____

Artwork Used: _____

Other Notes: _____

_____

_____

Tip: *Make sure your pets aren't in or near your studio if you don't want to record their voices as well.*

Podcast Name: _____

Recording Date: _____

Publication Date: _____

Notes of Relevance: _____

_____

_____

Host(s): _____

Guest(s): _____

Episode Number: _____ Date: _____

Episode Name: _____

Season Number: _____

Season Name: _____

RSS and podcast host/service provider: _____

_____

DAW: _____

Running Time: _____

Manual Uploads: _____

_____

_____

Categories: _____

Social Media Posts: _____

_____

Background Music: _____

Audio Clips: _____

Sponsorship and Placement: _____

_____

Metadata, ID Tags: _____

Artwork Used: _____

Other Notes: _____

_____

_____

Tip: *Create a brief noise profile as a first step in editing your recording.*

Podcast Name: _____

Recording Date: _____

Publication Date: _____

Notes of Relevance: _____

_____

_____

Host(s): _____

Guest(s): _____

Episode Number: _____ **Date:** _____

Episode Name: _____

Season Number: _____

Season Name: _____

RSS and podcast host/service provider: _____

_____

DAW: _____

Running Time: _____

Manual Uploads: _____

_____

_____

Categories: _____

Social Media Posts: _____

_____

Background Music: _____

Audio Clips: _____

Sponsorship and Placement: _____

_____

Metadata, ID Tags: _____

Artwork Used: _____

Other Notes: _____

_____

_____

Tip: *Warm up. Read and re-read. Exercise your mouth muscles.*

Podcast Name: _____

Recording Date: _____

Publication Date: _____

Notes of Relevance: _____

_____

_____

Host(s): _____

Guest(s): _____

Episode Number: _____ Date: _____

Episode Name: _____

Season Number: _____

Season Name: _____

RSS and podcast host/service provider: _____

_____

DAW: _____

Running Time: _____

Manual Uploads: _____

_____

Categories: _____

Social Media Posts: _____

_____

Background Music: _____

Audio Clips: _____

Sponsorship and Placement: _____

_____

Metadata, ID Tags: _____

Artwork Used: _____

Other Notes: _____

_____

_____

Tip: *Don't try to sound like a "radio professional". Your passion and content make your voice sound great.*

Podcast Name: _____

Recording Date: _____

Publication Date: _____

Notes of Relevance: _____

_____

_____

Host(s): _____

Guest(s): _____

Episode Number: _____ Date: _____

Episode Name: _____

Season Number: _____

Season Name: _____

RSS and podcast host/service provider: _____

_____

DAW: _____

Running Time: _____

Manual Uploads: _____

_____

_____

Categories: _____

Social Media Posts: _____

_____

Background Music: _____

Audio Clips: _____

Sponsorship and Placement: _____

_____

Metadata, ID Tags: _____

Artwork Used: _____

Other Notes: _____

_____

_____

Tip: *Try to be aware of how you use your hands when speaking. This can help avoid hitting the mic stand or other sensitive equipment in your studio.*

Podcast Name: _____

Recording Date: _____

Publication Date: _____

Notes of Relevance: _____

_____

_____

Host(s): _____

Guest(s): _____

Episode Number: _____ Date: _____

Episode Name: _____

Season Number: _____

Season Name: _____

RSS and podcast host/service provider: _____

_____

DAW: _____

Running Time: _____

Manual Uploads: _____

_____

_____

Categories: _____

Social Media Posts: _____

_____

Background Music: _____

Audio Clips: _____

Sponsorship and Placement: _____

_____

Metadata, ID Tags: _____

Artwork Used: _____

Other Notes: _____

_____

_____

2022 CALENDAR

JANUARY

Su	Mo	Tu	We	Th	Fr	Sa
26	27	28	29	30	31	1
2	3	4	5	6	7	8
9	10	11	12	13	14	15
16	17	18	19	20	21	22
23	24	25	26	27	28	29
30	31	1	2	3	4	5

FEBRUARY

Su	Mo	Tu	We	Th	Fr	Sa
30	31	1	2	3	4	5
6	7	8	9	10	11	12
13	14	15	16	17	18	19
20	21	22	23	24	25	26
27	28	1	2	3	4	5
6	7	8	9	10	11	12

MAY

Su	Mo	Tu	We	Th	Fr	Sa
1	2	3	4	5	6	7
8	9	10	11	12	13	14
15	16	17	18	19	20	21
22	23	24	25	26	27	28
29	30	31	1	2	3	4
5	6	7	8	9	10	11

JUNE

Su	Mo	Tu	We	Th	Fr	Sa
29	30	31	1	2	3	4
5	6	7	8	9	10	11
12	13	14	15	16	17	18
19	20	21	22	23	24	25
26	27	28	29	30	1	2
3	4	5	6	7	8	9

SEPTEMBER

Su	Mo	Tu	We	Th	Fr	Sa
28	29	30	31	1	2	3
4	5	6	7	8	9	10
11	12	13	14	15	16	17
18	19	20	21	22	23	24
25	26	27	28	29	30	1
2	3	4	5	6	7	8

OCTOBER

Su	Mo	Tu	We	Th	Fr	Sa
25	26	27	28	29	30	1
2	3	4	5	6	7	8
9	10	11	12	13	14	15
16	17	18	19	20	21	22
23	24	25	26	27	28	29
30	31	1	2	3	4	5

2022 CALENDAR

MARCH

Su	Mo	Tu	We	Th	Fr	Sa
27	28	1	2	3	4	5
6	7	8	9	10	11	12
13	14	15	16	17	18	19
20	21	22	23	24	25	26
27	28	29	30	31	1	2
3	4	5	6	7	8	9

APRIL

Su	Mo	Tu	We	Th	Fr	Sa
27	28	29	30	31	1	2
3	4	5	6	7	8	9
10	11	12	13	14	15	16
17	18	19	20	21	22	23
24	25	26	27	28	29	30
1	2	3	4	5	6	7

JULY

Su	Mo	Tu	We	Th	Fr	Sa
26	27	28	29	30	1	2
3	4	5	6	7	8	9
10	11	12	13	14	15	16
17	18	19	20	21	22	23
24	25	26	27	28	29	30
31	1	2	3	4	5	6

AUGUST

Su	Mo	Tu	We	Th	Fr	Sa
31	1	2	3	4	5	6
7	8	9	10	11	12	13
14	15	16	17	18	19	20
21	22	23	24	25	26	27
28	29	30	31	1	2	3
4	5	6	7	8	9	10

NOVEMBER

Su	Mo	Tu	We	Th	Fr	Sa
30	31	1	2	3	4	5
6	7	8	9	10	11	12
13	14	15	16	17	18	19
20	21	22	23	24	25	26
27	28	29	30	1	2	3
4	5	6	7	8	9	10

DECEMBER

Su	Mo	Tu	We	Th	Fr	Sa
27	28	29	30	1	2	3
4	5	6	7	8	9	10
11	12	13	14	15	16	17
18	19	20	21	22	23	24
25	26	27	28	29	30	31
1	2	3	4	5	6	7

2023 CALENDAR

JANUARY

Su	Mo	Tu	We	Th	Fr	Sa
1	2	3	4	5	6	7
8	9	10	11	12	13	14
15	16	17	18	19	20	21
22	23	24	25	26	27	28
29	30	31	1	2	3	4
5	6	7	8	9	10	11

FEBRUARY

Su	Mo	Tu	We	Th	Fr	Sa
29	30	31	1	2	3	4
5	6	7	8	9	10	11
12	13	14	15	16	17	18
19	20	21	22	23	24	25
26	27	28	1	2	3	4
5	6	7	8	9	10	11

MAY

Su	Mo	Tu	We	Th	Fr	Sa
30	1	2	3	4	5	6
7	8	9	10	11	12	13
14	15	16	17	18	19	20
21	22	23	24	25	26	27
28	29	30	31	1	2	3
4	5	6	7	8	9	10

JUNE

Su	Mo	Tu	We	Th	Fr	Sa
28	29	30	31	1	2	3
4	5	6	7	8	9	10
11	12	13	14	15	16	17
18	19	20	21	22	23	24
25	26	27	28	29	30	1
2	3	4	5	6	7	8

SEPTEMBER

Su	Mo	Tu	We	Th	Fr	Sa
27	28	29	30	31	1	2
3	4	5	6	7	8	9
10	11	12	13	14	15	16
17	18	19	20	21	22	23
24	25	26	27	28	29	30
1	2	3	4	5	6	7

OCTOBER

Su	Mo	Tu	We	Th	Fr	Sa
1	2	3	4	5	6	7
8	9	10	11	12	13	14
15	16	17	18	19	20	21
22	23	24	25	26	27	28
29	30	31	1	2	3	4
5	6	7	8	9	10	11

2023 CALENDAR

MARCH

Su	Mo	Tu	We	Th	Fr	Sa
26	27	28	1	2	3	4
5	6	7	8	9	10	11
12	13	14	15	16	17	18
19	20	21	22	23	24	25
26	27	28	29	30	31	1
2	3	4	5	6	7	8

APRIL

Su	Mo	Tu	We	Th	Fr	Sa
26	27	28	29	30	31	1
2	3	4	5	6	7	8
9	10	11	12	13	14	15
16	17	18	19	20	21	22
23	24	25	26	27	28	29
30	1	2	3	4	5	6

JULY

Su	Mo	Tu	We	Th	Fr	Sa
25	26	27	28	29	30	1
2	3	4	5	6	7	8
9	10	11	12	13	14	15
16	17	18	19	20	21	22
23	24	25	26	27	28	29
30	31	1	2	3	4	5

AUGUST

Su	Mo	Tu	We	Th	Fr	Sa
30	31	1	2	3	4	5
6	7	8	9	10	11	12
13	14	15	16	17	18	19
20	21	22	23	24	25	26
27	28	29	30	31	1	2
3	4	5	6	7	8	9

NOVEMBER

Su	Mo	Tu	We	Th	Fr	Sa
29	30	31	1	2	3	4
5	6	7	8	9	10	11
12	13	14	15	16	17	18
19	20	21	22	23	24	25
26	27	28	29	30	1	2
3	4	5	6	7	8	9

DECEMBER

Su	Mo	Tu	We	Th	Fr	Sa
26	27	28	29	30	1	2
3	4	5	6	7	8	9
10	11	12	13	14	15	16
17	18	19	20	21	22	23
24	25	26	27	28	29	30
31	1	2	3	4	5	6

2024 CALENDAR

JANUARY

Su	Mo	Tu	We	Th	Fr	Sa
31	1	2	3	4	5	6
7	8	9	10	11	12	13
14	15	16	17	18	19	20
21	22	23	24	25	26	27
28	29	30	31	1	2	3
4	5	6	7	8	9	10

FEBRUARY

Su	Mo	Tu	We	Th	Fr	Sa
28	29	30	31	1	2	3
4	5	6	7	8	9	10
11	12	13	14	15	16	17
18	19	20	21	22	23	24
25	26	27	28	29	1	2
3	4	5	6	7	8	9

MAY

Su	Mo	Tu	We	Th	Fr	Sa
28	29	30	1	2	3	4
5	6	7	8	9	10	11
12	13	14	15	16	17	18
19	20	21	22	23	24	25
26	27	28	29	30	31	1
2	3	4	5	6	7	8

JUNE

Su	Mo	Tu	We	Th	Fr	Sa
26	27	28	29	30	31	1
2	3	4	5	6	7	8
9	10	11	12	13	14	15
16	17	18	19	20	21	22
23	24	25	26	27	28	29
30	1	2	3	4	5	6

SEPTEMBER

Su	Mo	Tu	We	Th	Fr	Sa
1	2	3	4	5	6	7
8	9	10	11	12	13	14
15	16	17	18	19	20	21
22	23	24	25	26	27	28
29	30	1	2	3	4	5
6	7	8	9	10	11	12

OCTOBER

Su	Mo	Tu	We	Th	Fr	Sa
29	30	1	2	3	4	5
6	7	8	9	10	11	12
13	14	15	16	17	18	19
20	21	22	23	24	25	26
27	28	29	30	31	1	2
3	4	5	6	7	8	9

2024 CALENDAR

MARCH

Su	Mo	Tu	We	Th	Fr	Sa
25	26	27	28	29	1	2
3	4	5	6	7	8	9
10	11	12	13	14	15	16
17	18	19	20	21	22	23
24	25	26	27	28	29	30
31	1	2	3	4	5	6

APRIL

Su	Mo	Tu	We	Th	Fr	Sa
31	1	2	3	4	5	6
7	8	9	10	11	12	13
14	15	16	17	18	19	20
21	22	23	24	25	26	27
28	29	30	1	2	3	4
5	6	7	8	9	10	11

JULY

Su	Mo	Tu	We	Th	Fr	Sa
30	1	2	3	4	5	6
7	8	9	10	11	12	13
14	15	16	17	18	19	20
21	22	23	24	25	26	27
28	29	30	31	1	2	3
4	5	6	7	8	9	10

AUGUST

Su	Mo	Tu	We	Th	Fr	Sa
28	29	30	31	1	2	3
4	5	6	7	8	9	10
11	12	13	14	15	16	17
18	19	20	21	22	23	24
25	26	27	28	29	30	31
1	2	3	4	5	6	7

NOVEMBER

Su	Mo	Tu	We	Th	Fr	Sa
27	28	29	30	31	1	2
3	4	5	6	7	8	9
10	11	12	13	14	15	16
17	18	19	20	21	22	23
24	25	26	27	28	29	30
1	2	3	4	5	6	7

DECEMBER

Su	Mo	Tu	We	Th	Fr	Sa
1	2	3	4	5	6	7
8	9	10	11	12	13	14
15	16	17	18	19	20	21
22	23	24	25	26	27	28
29	30	31	1	2	3	4
5	6	7	8	9	10	11

PINPOINT YOUR LISTENERS

A MAP OF YOUR COMMUNITY

Get Your Copy of *Podcasting – Explained!* at an Amazon Global Marketplace today!

www.henleypoint.ca

STEVEN CHRISTIANSON

Henley Point Productions 2022

ABOUT THE AUTHOR

Steven Christianson's career experience spans Canada's Parliament and the United Nations, the corporate and non-profit sectors. For several years he taught advocacy as an instructor with Ryerson University (now Toronto Metropolitan University). A public policy analyst by vocation, and an avid podcaster and content creator, his work is defined by developing content, promoting it, and making it work.

Steven's professional career has focused on government relations, advocacy, communications events and political management.

Steven began corporate webcasting in the mid-1990s, and published his first episodic podcast in 2005.

Steven holds a Master's degree in public policy and political economy. He is the author of a special edition book, *Canada's Indian Act: Policy Perspectives from the Years Defined by Oka, Meech Lake and the Royal Commission*, as well as the *Explained!* series, titles which include *Content Creation and Promotion*, *Advocacy* and *Podcasting*, as well as two companion readers, *Advocacy – The Companion Handbook: A Guide to Advocacy and Lobbying in Canada and the United States,* and *Podcasting – The Companion Handbook: A Guide to Producing and Publishing Your Podcast.*

Steven has one daughter, Courtney, and lives in Toronto with his wife, Josée, and two cats, Sophie and Jules.